To Cath, with love. CS

To the tree keepers: Libby, Nic, Sander, Stu and Victoria; and to Kaz Brown: thank you. JR

First published by Allen & Unwin in 2024

Copyright © Text, Claire Saxby 2024
Copyright © Illustrations, Jess Racklyeft 2024

All rights reserved. No part of this book may be reproduced or transmitted in any form or by any means, electronic or mechanical, including photocopying, recording or by any information storage and retrieval system, without prior permission in writing from the publisher. The Australian *Copyright Act 1968* (the Act) allows a maximum of one chapter or ten per cent of this book, whichever is the greater, to be photocopied by any educational institution for its educational purposes provided that the educational institution (or body that administers it) has given a remuneration notice to the Copyright Agency (Australia) under the Act.

Allen & Unwin
Cammeraygal Country
83 Alexander Street
Crows Nest NSW 2065
Australia
Phone: (61 2) 8425 0100
Email: info@allenandunwin.com
Web: www.allenandunwin.com

*Allen & Unwin acknowledges the Traditional Owners of the Country on which we live and work.
We pay our respects to all Aboriginal and Torres Strait Islander Elders, past and present.*

 A catalogue record for this book is available from the National Library of Australia

ISBN 978 1 76106 950 5

For teaching resources, explore allenandunwin.com/learn

Illustration technique: watercolour, acrylic painting, collage, pencil, ink and digital illustration

Cover and text design by Sandra Nobes
Set in 18 pt Century OS MT Std, hand lettering by Jess Racklyeft
This book was printed in November 2023 by C&C Offset Printing Co. Ltd, China

1 3 5 7 9 10 8 6 4 2

www.clairesaxby.com
www.jessesmess.com

TREE

Claire Saxby • Jess Racklyeft

ALLEN&UNWIN
SYDNEY•MELBOURNE•AUCKLAND•LONDON

Can you see the forest on this misty-morning mountain?

Can you see where the tree stands?

It is the tallest in this forest of tall trees.

This tree is older than those who find it,
younger than the land it grows from.
Every day, its roots drink in water.
Every day, its leaves use light to make energy.
Between sapwood and heartwood, water rises and energy flows.
The tree grows.

See the mist melt and bright the sky blue.

See the branches lean, so far above.

Listen to the leaves bustle. Smell the forest air.

This tree breathes in the air we breathe out,

    breathes out the air we breathe in.

This is the world of the tree.

Beneath the ground, white threads fine as floss connect this tree to others.

In the layered litter, a scaly thrush flicks.
A lyrebird scritch-scratches.
Slaters curl, beetles burrow
and centipedes scurry.

A robin perches on a wattle branch,
watches and waits.
Beneath the bracken, a fairy-wren chitters
as a spotted pardalote disappears
into the tunnel that leads to her nest.

Look carefully.
There – on the mossy log –
a blue-tongued lizard
basks in a sunpatch.

An echidna noses under the log,
searching for ants.
The lizard vanishes and the ants scatter,
track up the mossy bark of the tree.
Between sapwood and heartwood,
water rises and energy flows.

A rainstorm begins. Ends.

Drops pool where fronds unfurl in a tree fern crown.

Hold out your hand, catch a raindrip.

Rain feeds the tree, feeds the forest.

A breeze stumbles about and a silver wattle shivers.

Wattles thrive in the dapple of the tree.

A white-throated treecreeper pee-peets as it spirals upward,
   past the hollow where a cuddle
   of Leadbeater's possums sleep.
The treecreeper finds a beetle in the bark.
In the late afternoon, a pair of rosellas alight
   in the high branches,
pick insects from the leaves and travel on.

Day meets night and a greater glider emerges from her hollow. She grooms herself then launches into the dusk.

Close your eyes. Listen to the new sounds of night.
The tiny possums are awake now, alert for owls.
One-by-one they leap from their nest hollow.
Later they will shred more stringy bark to
 re-cosy their nest.

A yellow-bellied glider climbs to where the bark is thin.
He bites through it, laps the sugary sap.
He startles at the tu-whoo of a powerful owl. Hides.
When the owl calls again, she is softer, further away.
Now it is safe for the glider. He sails to a younger tree,
	makes an impossible landing.

A mountain brushtail possum flits from wattle to wattle nibbling leaves. Branches bounce and her large night-eyes glint in moonlight.

Down on the ground, a boulder-big wombat tears at tufty grass. Chews. Wallabies thump by and the wombat barrels away through the undergrowth. A bush rat digs for fungi in the crevices between moss-covered buttresses. A boobook owl swoops.

Between sapwood and heartwood,
water rises and energy flows.

The tree grows.

One day this tree will fall and become part of the forest floor.

Moss will grow.

Echidnas will nose beneath it.

Lizards will shelter inside it.

But not today.

Today the tree stands tall.

Another day dawns.